PENGUINS
CLOSE ENCOUNTERS

First published in 2013 by New Holland Publishers
This edition published in 2014 by Bloomsbury Publishing

Bloomsbury Publishing Plc, 50 Bedford Square, London WC1B 3DP

www.bloomsbury.com

Bloomsbury is a trademark of Bloomsbury Publishing Plc

Bloomsbury Publishing, London, New Delhi, New York and Sydney

A CIP catalogue record for this book is available from the British Library.

Publisher: Nigel Redman
Designer: Lorena Susak

ISBN 978 -1-4729-1000-4

Printed and bound in China by Toppan Leefung Printing Ltd

This book is produced using paper that is made from wood grown in managed sustainable forests. It is natural, renewable and recyclable. The logging and manufacturing processes conform to the environmental regulation of the country of origin.

All images by David Tipling except the following: Nigel McCall (page 153), Pete Morris (pages 33, 148, 149, 152, 155 and 156 both images), Nature PL (pages 54, 124 and 158) and Brent Stephenson (page 151).

PENGUINS
CLOSE ENCOUNTERS

DAVID TIPLING

B L O O M S B U R Y
LONDON • NEW DELHI • NEW YORK • SYDNEY

Contents

Introduction

Never have I felt so alive in spirit or so vulnerable to death as the day I photographed Emperor Penguins during an Antarctic storm. It was 11 November 1998 and I had spent 10 days camping on the sea ice in the Weddell Sea, not far from the very spot where Lord Shackleton had abandoned his ship the *Endurance* in 1915. A storm had swept across from the west and, with icy winds in excess of 80kph (50mph), it had blown up into a raging white out. The sea ice groaned and, more worryingly still, it was beginning to crack. A few hours earlier I had sat in my tent clinging to the roof while it lurched violently in the strong gusts. Yet, as conditions improved a little, I made it back to the Emperor Penguin colony in which I had immersed myself for days. I had photographed the life of a bird that is, for me at least, the ultimate. No other warm-blooded animal on Earth endures such intense cold or has such an extraordinary breeding strategy.

In that vile hurricane I stood very close to the Emperors – at least 200 chicks huddled together in a crèche surrounded by their braying parents. I worked quickly to create more images, but somehow photography seemed secondary in this incredible moment. Somewhere out there was the rest of the expedition team with whom I'd travelled to this remarkable place, but they were all invisible to me in the endless blast of snow. I was alone with these birds, and it felt special, as if I were living another life, cut off from the world I had left far behind. Even as I stood there amid all that snow and surrounded by all these sheltering birds, I realized that I was witness to a spectacle and party to a moment coming but once or twice in any lifetime.

It was also hard to believe that just ten days before this wonderful intimacy with the Emperors among such savage beauty, I had had my first ever sighting of a penguin. In truth it could not have been a more different experience, nor a more remote kind of encounter. It came from the seat of the aircraft in which we flew to the Emperor colony. At 3,700m (12,000ft) elevation, a group of what may have been Chinstrap or Adélie Penguins looked like mere atoms of darkness scattered across a vast iceberg afloat in the Southern Ocean. I was peering out of the cockpit window and below me a land of incredible beauty was slowly unfolding. We were on a six-hour flight from Punta Arenas in Chile to Patriot Hills, a summer camp at the base of the Ellsworth Mountains situated at 80°S and just 966km (600 miles) from the South

Pole. From here I would be flying another 1,100km (700 miles) to the Dawson-Lambton Glacier.

Our wheeled Hercules touched down on the blue ice runway at Patriot Hills and was described by our pilot as 'real seat-of-the-pants flying.' If I'm honest, by the time we came to a halt I was in need of a change of underwear! The aircraft slewed and snaked across the ice. Then as I walked down the ramp of the Hercules the cold was brutal. For the next three weeks I was going to be camping on that ice. I remember thinking at the time: it'll be like living in a chest freezer. When I washed my hair in a bowl of steaming water it froze immediately afterwards into hard icy spikes. When katabatic winds roared down at us from the South Pole bringing hurricane-force gales, any flesh exposed even for a few seconds would be frostbitten.

Patriot Hills was home as the storms swept across us, but when a break in the weather came a few days later we set off finally for the Dawson-Lambton Glacier. As we flew over the frozen Weddell Sea, the view below was exhilarating. Massive tabular icebergs sat within a white plane of pack ice and towered over row upon row of enormous deep-blue pressure ridges, whose crystalline edges sparkled in the brilliant sunlight. Yet the strongest memories from that day in early November 1998 are of my first real penguin encounter.

We had landed on the sea ice and set up camp and had then walked in the early hours of the morning around 1.5km (1 mile) along the edge of a towering ice-cliff to an Emperor Penguin colony. An extract from my diary recounts the experience:

'Still out of sight but just a few steps away the sounds of adults trumpeting and chicks peeping quickened our stride. Then ahead of us were thousands of Emperors clustered together in seven sub-colonies within an amphitheatre of sculpted ice, the low light washing a warm glow over the birds. I soon got to work making many pictures. By 8am I felt exhausted and retired to camp for bread and fried eggs. I slept until 4pm and after a quick snack went back to the penguins, where after a short search I found a tiny chick being brooded on its parent's feet. I sat for many hours in this corner of the colony, often adults and chicks approached touching my boots with their beaks.'

That first encounter with penguins had me hooked. Since then I have peered into sea caves at fluffy Little Penguin chicks in New Zealand, strolled with Gentoo Penguins along the sandy beaches in the Falklands and stood and marvelled at King Penguin colonies that stretch for as far as the eye can see.

I realize that I am not alone in having such deep feelings for these birds. My passion is shared by people all over the world. Testament to this is the huge success of films like *March of the Penguins* (2005) and *Happy Feet* (2006). The first of these is the second highest grossing documentary at the box office. It was created at the Emperor Penguin colony at Dumont

d'Urville, in the Antarctic region known as Adélie Land, and follows the birds' complete breeding cycle. The anthropomorphic qualities of this remarkable species, with its waddling upright gait and arm-like flippers, easily translated into a human story about love, family values and survival.

Given that penguins live in some of the remotest parts of our planet, they are often used as symbols of wilderness. It was an attribute that did not go unnoticed by the early explorers, who often annotated maps using penguin colonies as landmarks or even place names. These were not noted down for their tourist interest, of course, but for their survival value as sources of fresh protein. In 1578 the English explorer Sir Francis Drake described the taste of Magellanic Penguin flesh as not unlike a fat goose in England. In the 19th century, sealers turned their attention to slaughtering penguins. On the Falkland Islands an estimated 1.5 million Rockhopper Penguins were boiled down for their oil in a little more than a decade, while similar wholesale harvests decimated colonies across the southern hemisphere. A low-level slaughter of penguins still occurs today, but there are now far greater risks to the birds. These new threats include the impacts of climate change and El Niño events, over-fishing, oil spills, and even the effects of scientists or tourists at some heavily visited penguin colonies.

Despite our dismal treatment of them in the past, wild penguins almost everywhere have not lost their deep, abiding sense of curiosity towards humans. Regularly the wild birds will wander right up to people and inspect them as if they were just another kind of flightless creature walking on two legs. Their extraordinary antics have embedded penguins in our stories and fables; they have elevated them as heroes in films and documentaries; they have given them an allure that draws thousands of people to the extraordinary and remote places where penguins live. These are birds that never fail to make people smile.

With the exception of the Galápagos Penguin, whose range straddles the Equator, the family is confined to the southern hemisphere. All of the species are equipped for life in cold water. A torpedo shape, flat flippers in place of feathered wings and eyes adjusted for underwater vision are adaptations to a life spent largely in the sea. Penguin plumage offers perfect camouflage, with a white belly to disguise the bird from underwater attack, and a dark back to hide it from predators that lurk overhead.

The name 'penguin' means 'fat' in Spanish and probably comes from a time when Hispanic sailors hunted for another rotund and flightless seabird of the northern hemisphere, the Great Auk. When these same mariners then travelled to the southern oceans and met different kinds of fleshy fowl, they called these new birds penguins too. The layer of blubber in a penguin's body acts as an energy store during the lean times when the birds cannot easily obtain food. Classic examples of these fasts include the months spent by parents incubating eggs and caring for their young, or during the period when an adult penguin moults its feathers and is confined to land. Blubber also conducts heat poorly so it acts as insulation to maintain the bird's core body temperature. This is particularly vital for a species such as the Emperor Penguin, which endures some of the most brutal weather on our planet.

This book is a visual celebration of a family of birds that have given me more pleasure photographing and observing than any other. Each chapter explores in words and pictures an aspect of penguins' lives, with a final chapter that gives a detailed review of each species. I make no apology for including more images from both Antarctica and subantarctic islands, because it is where I have spent most of my time with these remarkable birds. I have lived among Emperor Penguins which, in my biased view, is the most remarkable bird on the planet. Ironically, this most remote of all species is probably the penguin with which more people are familiar. This is partly because it epitomizes the penguin stereotype – the flightless giant that dwells amid ice and snow and rears the cutest, most endearing chick in all the world.

The book ends with a brief overview of each of the penguin species. Bird taxonomy is in a state of constant flux as we learn more about how members of a particular bird family relate to one another and to their closest relatives. Some authorities now recognize up to 22 separate penguin species, but I adhere to a family total of 17, because this is the figure recognized in most popular literature including field guides.

I hope this book will be enjoyed by all who love penguins, regardless of whether you have been privileged to meet the birds personally in the wild, or have simply enjoyed their antics at the cinema, or on the television screen, or in the zoo. I also hope my pictures will inform you and that they will, perhaps, bring back memories of your own encounters with penguins. Above all, however, I hope they will make you smile.

David Tipling

Penguin land

Gentoo Penguins stand below the hanging glaciers and snow-capped peaks that loom over Gold Harbour on the South Atlantic Island of South Georgia. At least 300 pairs of this species nest at the site in the tussock grass behind the beach, but to reach them one must first negotiate a wall of blubber created by the breeding rookery of Southern Elephant Seals. Few places on Earth support a similar abundance of wildlife in such a spectacular setting and initially it can be almost overwhelming. To take this shot I took advantage of the long sunny summer days, catching the crystalline light just after 4am.

A small group of Gentoo Penguins sits out a blizzard at the decaying whaling station of Grytviken on South Georgia. Towering over them is the rusting hulk of *Petrel*, an old whaling vessel.

PETRE

Gold Harbour on South Georgia is home to 50,000 pairs of King Penguins but there are more than 1 million pairs of this species breeding on subantarctic and Antarctic islands. Getting right down to take the photograph at penguin level helps to create a more intimate feel to this adult as he trumpets in courtship from the middle of a packed colony.

Macaroni Penguins favour nesting sites on steep, often inaccessible, hillsides. After a precipitous scramble up a scree slope and then through thick tussock grass, I reached a small colony of Macaronis at Hercules Bay on South Georgia. It was early in the breeding season and snow still lay on the ground. To create this unusual angle I mounted my camera at the end of a long pole and fixed a wide-angle lens, firing the shutter with a remote trigger.

The Emperor Penguin thrives in a land that is the darkest, driest, coldest and windiest place on Earth. I lay on the ground and used a panoramic camera to capture this solitary individual within the vast white emptiness that is the species' remarkable home.

Chinstrap Penguins are highly gregarious and often gather in their hundreds, and sometimes thousands, on the immense icebergs drifting between the subantarctic islands and the Antarctic continent.

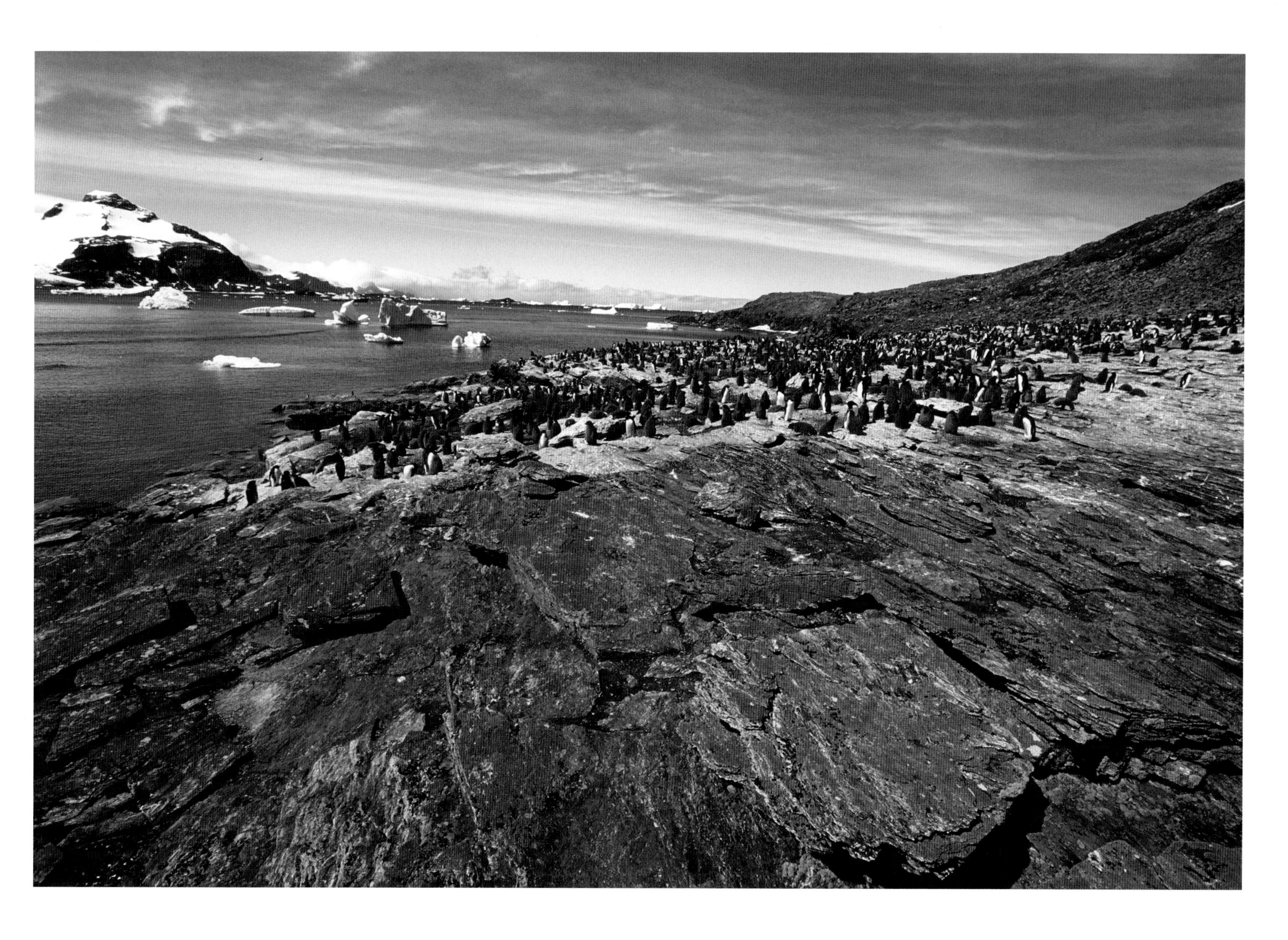

The glorious lichens that encrust the rocks at this Adélie Penguin colony at Shingle Cove on Livingston Island add an unexpected splash of almost tropical colour to the scene.

Commuter routes crisscross the otherwise dense tussock grass that smothers the hillside at this Macaroni Penguin colony in Cooper Bay, South Georgia. I lay down beside one of these tracks and photographed inquisitive individuals as they stopped momentarily to investigate my prostrate figure. A fish-eye lens gives that lovely wide view of the bird in its specialized habitat.

A thin dark line of life in a vast desert of white – Emperor Penguins toboggan for 50km (30 miles) across the sea-ice from their sheltered breeding grounds to the open ocean. They will make this trek many times, but as the ice retreats during the summer each of their journeys will become shorter than the last.

The King Penguin colony at St Andrews Bay, South Georgia, is one of the greatest wildlife spectacles on Earth. In summer nearly 200,000 pairs breed on the boulder-strewn plain of glacial outwash, which slowly sluices down from three receding glaciers nearby. On my first visit I remember putting my camera away for a few minutes, while I simply stood and looked in order to take in the full magnificence of what lay before me.

At Brown Bluff on the Antarctic Peninsula, a small group of Gentoo Penguins departs on a fishing trip. In common with other Antarctic family members, Gentoo Penguins feed predominantly on small shrimp-like crustaceans, known as Antarctic Krill, as well as on fish and squid. Here the birds pass a small iceberg fragment that is known as a 'bergy bit'.

At Half Moon Island, just off the Antarctic Peninsula, an impressive column of basalt had thrust through the ice to create a pocket of dry land amid the snow. Chinstrap Penguins crowded on to this spot in order to breed. The Antarctic summer is relatively short and the birds try to time the fledging of their young to coincide with the maximum prey abundance in the Southern Ocean.

I was captivated by the play of light and shadow on the vast expanse of snow that lay beyond this low ice cliff. I had a hunch that if I watched and waited, then a passing penguin or two would bring a sense of scale and an element of foreground interest to this powerful scene. I sat for nearly two hours until a line of four Emperors walked into frame. My picture was complete.

The approachability of penguins allows the photographer to experiment with angle and composition. At St Andrews Bay on South Georgia I lay on the ground and photographed just the heads and shoulders of King Penguins as they wandered by. I used a 10.5mm (0.4in) fish-eye lens that distorted the line of the horizon but it also helped to emphasize the sky.

My eye was drawn to the rich contrasts of light and dark on this volcanic beach in South Georgia. The colour of the sand somehow complimented, yet also highlighted, the tones on the King Penguins that had ambled down to the shoreline. The birds seemed hesitant, perhaps on the look out for Killer Whales which had been patrolling this beach just a few hours earlier.

A vast boulder-field of blubber blocks the way for King Penguins as they head back and forth to the sea. In summer nearly 6,000 Southern Elephant Seals haul out at St Andrews Bay, South Georgia. With some of them weighing the equivalent of a large car, the seals fight, court, mate and give birth at this spot. Unfortunately for the King Penguins their presence creates an ever-moving obstacle course.

Pressure ridges rise from the Weddell Sea at a spot close to the Dawson-Lambton Glacier. While they were structures of great beauty to my photographic eye, these walls of ice were, in fact, major obstacles to the Emperor Penguins, which must find a way through them to reach their breeding grounds. As I explored the surrounding area I found some of these ridges a serious challenge even for me!

Young King Penguins cluster together in great crèches that involve hundreds, sometimes thousands, of birds. Each parent can identify its own offspring by its voice, a feat that seems all the more remarkable when you stand among such a crowd and listen to their incessant chirping.

A colourful adult King Penguin at the St Andrews Bay colony, South Georgia, contrasts against a backdrop of downy brown young birds.

Royal Penguins packed shoulder-to-shoulder on Macquarie Island, which is home to the world's largest colony holding 500,000 pairs.

On the move

Although penguins are best adapted to their life in the water, they are no mean movers on dry land. I came across this group of Gentoo Penguins running excitedly down a Falkland Island beach. They were travelling at such a pace I would have had to run hard if I'd wanted to keep up.

Once across the boulders, the Rockhopper Penguins at many colonies have to climb up steep slopes and follow routes worn down by centuries of constant use. This trio of birds on the Falkland Islands were making their way back to their nests after a fishing trip – a journey that will take them more than an hour as they traverse various ledges and at least one steep cliff.

Known as 'Jumping Jacks' on the Falkland Islands, Rockhopper Penguins can easily hop from boulder to boulder and make much the same body movements as we do when we're competing in a sack race.

Marching in single file, these Emperor Penguins are heading for their colony in the lee of a cliff at the edge of Antarctica's Weddell Sea. They have already walked nearly 50 miles from the edge of the pack ice and when they finally get back to their respective mates, the chicks will receive their first meal in several days.

In summer the sun never sets at this latitude, so during my stay at this Emperor Penguin colony I decided to sleep during the day and walk with the penguins at night. It enabled me to take advantage of the softer, warmer light as the sun reached its lowest angle. I took this particular picture at 3am.

On the same day that I took this picture on South Georgia, my diary entry reads:

'I spent a few hours today sitting on the beach waiting for King Penguins to emerge from the surging surf, their explosive arrival was followed by a walk reminiscent of Charlie Chaplin's famous waddle. I could not stop smiling at each performance.'

It was once suggested to Chaplin that he had derived his walk from watching penguins. Not so, he replied, it came from an old drunk he had once seen when he was a child in England. Nevertheless, the similarity between his flat-footed stroll and the gait of penguins has often been noted.

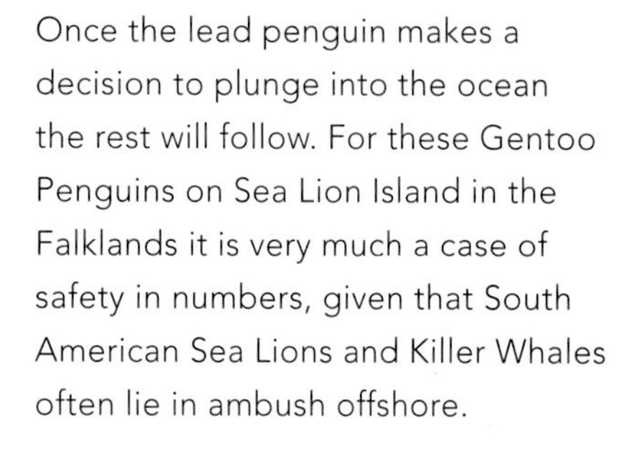

Once the lead penguin makes a decision to plunge into the ocean the rest will follow. For these Gentoo Penguins on Sea Lion Island in the Falklands it is very much a case of safety in numbers, given that South American Sea Lions and Killer Whales often lie in ambush offshore.

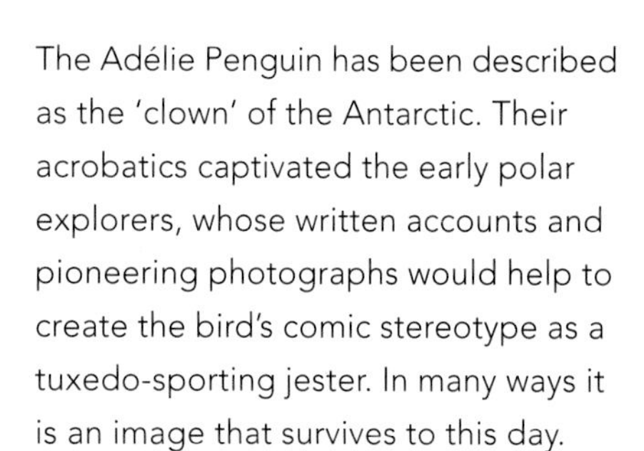

The Adélie Penguin has been described as the 'clown' of the Antarctic. Their acrobatics captivated the early polar explorers, whose written accounts and pioneering photographs would help to create the bird's comic stereotype as a tuxedo-sporting jester. In many ways it is an image that survives to this day.

A group of Adélie Penguins gingerly makes its way across a trench in an iceberg. Apsley Cherry-Garrard, a surviving member of Scott's fateful journey to the South Pole, noted in his book *The Worst Journey in the World*: 'They are extraordinarily like children, these little people of the Antarctic world, either like children or like old men, full of their own importance.'

There is an intriguing rule of polar photography: extreme weather equals extreme pictures. Generally the worse the weather gets, the more spectacular and impressive the images can be. During a landing at Right Whale Bay on South Georgia in early November, katabatic winds (cold winds that rush down from the glaciers and ice fields above the bay) suddenly created a blizzard. The wind was so strong that it was difficult for me to stand upright. The King Penguins formed a great long line as they walked to the lee of a mountain in search of shelter. Every few minutes I had to empty my lens hood as the swirling snow blew in my face, but this inconvenience and the numbing cold were both completely forgotten as I worked to capture the remarkable scenes unfolding before me.

A group of Adélie Penguins sets off on a fishing trip. The simple plumage of black back and white belly helps to disguise the penguin from predators once it is in the water.

As the polar winter draws near, Emperor Penguins leave the comfort of their ocean home to embark on a remarkable trek. The journey may take them a week or more as they travel in long lines, day and night, sometimes walking 110km (70 miles) across the sea ice to traditional breeding colonies. Thousands of generations before them have made the same march and all return to the place where they were born. Here they hope to find a mate and, as winter edges ever closer, the females will lay their eggs. Yet it is the male penguins that remain to incubate them through the Antarctic winter. Their partners then make the long journey back to the sea, not returning to the colony until the chicks are born.

A snaking crack in the sea ice, known as a 'lead', offers the birds an entry and exit point for the ocean below, where these Emperors – looking like mere points of darkness in a vast plain of white – will hunt for fish, squid and krill to feed their hungry chicks. This image was taken from a Russian helicopter hovering above the Weddell Sea near Signy Island in Antarctica.

A Gentoo Penguin surfaces. Birds that fly possess hollow bones to reduce their weight, but penguins have heavy solid bones that aid their sub-aquatic feeding habits. Large species, such as King and Emperor Penguins, can swim at an average speed of 14.5kph (9mph), while their medium-sized relatives, including Gentoo Penguin, can sustain 8kph (5mph) for hundreds of kilometres. The birds can also manage short bursts of 40-48kph (25-30mph) when hunting prey or when pursued by a predator. Penguins change the shape of their bodies to suit their swimming speed and steer by pressing their feet close to their tail.

A Gentoo Penguin rockets ashore on a Falkland Island beach, having run a potential gauntlet of predatory Killer Whales and South American Sea Lions.

A group of Magellanic Penguins on the Falkland Islands is heading for the sea. These birds are great ocean travellers, some reaching as far north as southern Brazil every year. One individual was recorded to swim 2,676km (1,663 miles) in 75 days.

I've spent many hours trying to photograph penguins as they come ashore. It is a challenge to find the perfect position from which to capture them at the exact moment that they leave the surf. A fast shutter speed is essential to freeze the action, 1/1000th sec in the case of this King Penguin that surfed a wave for the last few feet to the beach.

Perhaps just two or three times a year I capture an image that really excites me, a picture that transcends the customary portrait or action shot. This is one such photograph. I took it from the bow of the Russian icebreaker the *Kapitan Khlebnikov* at lunchtime. As I was eating I peered out of the window and noticed unusually calm conditions as we neared Brown Bluff on the Antarctic Peninsula. So I abandoned my meal and when I got to the bow a small group of Adélie Penguins were porpoising torpedo-like from below. Just this one image was sharp, with the penguin and its reflection perfectly framed.

Symmetry in an image can be a powerful ingredient. The anthropomorphic associations we have with penguins are never far away, and one could easily imagine that these Gentoo Penguins were following a dance routine or doing a keep-fit workout!

A Gentoo Penguin is such a graceful animal at sea, when its feet serve as a really effective rudder through the water. But on land the same extremities are supported by such short legs that the birds use twice as much energy while walking as any other animal of comparable height. Yet the penguin waddle is not just an outcome of all this inefficiency: scientists have concluded that their swaying action actually conserves energy and provides momentum.

When leaving the ocean Emperors power out of the water and land on the ice with a thud. They have thick fatty blubber that not only insulates them against the cold, but acts as a shock absorber when landing on their bellies. This bird is using a lead that opened up close to its colony in the Weddell Sea.

Under the water Emperor Penguins are transformed and their torpedo shape helps them to glide swiftly and effortlessly through the water. Here a flock is setting off on a fishing trip. Emperor Penguins possess an extraordinary ability to dive deeper and to swim underwater for longer than any other bird. They regularly dive for minutes at a time, but some birds have been recorded to stay under for more than 20 minutes. Emperors frequently fish at depths of 50–150m (164–492ft), but have been recorded going deeper than 400m (1,300ft), which is more than the height of the Empire State Building in New York.

It could be a scene from the film animation *Happy Feet*. Young Emperors around eight weeks old chase each other across the sea ice. At times I struggled to stay upright, so I was reassured that these birds, which are otherwise so adapted to a life on ice, could also find things a bit slippery.

Summer in an Emperor colony. Those chicks that are old enough to be left by themselves often go exploring. They reminded me of packs of teenagers hanging around on the edge of town.

Courtship

Macaroni Penguins have long-lasting pair bonds. Both birds engage each spring at their established nest site within a colony, recognizing each other by call. Cooper Bay on South Georgia supports a large Macaroni colony. Snow lay among the tussock grass and the birds busied themselves cementing their pair bonds and nest building. Here a male is in display, bowing before he stands upright and throws his head back to call in what is known as an 'ecstatic' display.

In the 1700s stylish Englishmen wore outlandish feathered headgear. Exceeding the ordinary bounds of fashion these foppish dandies became known as Macaronis. When English sailors came across this penguin in the Falklands with its punk-rocker hairstyle it reminded them of Macaronis back home, hence the name Macaroni Penguin.

Flippers back, calling skyward, this Chinstrap Penguin is performing the 'ecstatic' display. He is advertising his presence early in the season, laying claim to his nesting territory and advertising his vigour to any prospective mate.

Male Adélie Penguins arrive at the breeding colony a few days before females and start nest building with small stones. They are notorious thieves and take any opportunity to pinch stones from their neighbours. This colony on the Antarctic Peninsula seemed a microcosm of human society: played out on a daily basis were quarrels, violence, divorce, devotion, tough parenting and so much more.

This incubating Adélie Penguin along with its egg and nest have almost all disappeared under a blanket of early spring snow.

The golden-yellow ear and throat patches possessed by King Penguins are thought to be critical signals in courtship. Recent research suggests the colour intensity of the beak denotes both experience and age. These indicators of maturity help birds to choose an experienced partner and provide the best chance of breeding success. In this photograph I opted to use a very shallow depth of field, to place emphasis on the glorious colour and its shape.

A Magellanic Penguin incubates at the entrance to a burrow in early spring on the Falkland Islands. In places where Magellanic Penguins are unable to dig a burrow they lay their two eggs in a simple scrape often under a bush.

Like Adélie Penguins, Gentoo Penguins are notorious stone rustlers. A bigger nest may have a direct influence on the reproductive success of these birds. Nests with more stones are better protected from floods of snow-melt in early spring. Thomas Bagshawe studied Gentoo nesting behaviour in the early 1920s and summed it up in his book *Two Men in the Antarctic. An Expedition to Graham Land* where he wrote, 'Thieving is part of a penguin's nature and is so universally indulged in that it may be regarded as a normal habit rather than a sin.'

Mutual preening is a part of courtship for Rockhopper Penguins and is a way of removing ticks and fleas that can plague penguins when they have to spend long periods on land, either sitting on a nest or during their moult.

Half Moon Bay on the Antarctic Peninsula offers great penguin photography. On a cold clear evening as I made my way back down to the beach to take a zodiac back to the ship I came across this pair silhouetted against the sea. I was drawn to the backdrop of glistening water and the stance of these Gentoos, which conveyed a sense of contentment.

Good balance is a prerequisite for being a good lover if you are a penguin. This male Gentoo Penguin, in common with all its family members, does not possess a penis, so he has to line up his own cloaca with hers in order for sperm to be properly exchanged.

In high spring the bray and trumpet of displaying male Gentoos fills the air at rookeries of this particular penguin. Here is a bird in full cry.

Emperor Penguins find their soul mates by song. The birds also rely on sound rather than sight to locate both chick and partner. In his quest for a mate, a male Emperor will droop his head before raising it to trumpet loudly. It is a signal of his readiness to breed and if a female trumpets back, then they will come together to explore and cement their future partnership. Note the barbs inside this penguins beak used for gripping prey.

Gentoo Penguins use not only stones to make their nests, but also incorporate grass and soil. I sat out a blizzard at this Gentoo colony on Sea Lion Island on the Falklands, and while the incubating birds sat tight on their eggs, nest building continued for their partners.

Although these King Penguins allowed an approach to within touching distance, it was advantageous to use a telephoto lens to create this image. By using a shallow depth of field I isolated the front bird and positioned the two out-of-focus heads in the background, to create a composition that would draw the eye into the picture.

This male King Penguin is trying to impress with his sky-pointing display. I used a telephoto lens and shallow depth of field to create this portrait against a backdrop of out of focus chicks.

Four King Penguins stand in shadow, silhouetted against an ice-covered cliff at Royal Bay on South Georgia.

This pair of images illustrates two of the various postures taken up by a pair of Gentoo Penguins as they display at their newly established nest site.

Aggression is never far away in a Gentoo colony. Pairs nest close together but just out of reach of attack from their neighbours. This gesture is mere bluster.

A braying Gentoo Penguin in 'ecstatic' display shows off on the edge of the colony.

Coming across this trio of King Penguins, I was initially attracted by the anthropomorphic feel to the scene. Photographing wildlife with a commercial eye usually involves an ability to spot and shoot pictures that are generic, but which also suggest an underlying idea. In this case it is the notion of communication although, in truth, they are actually in an aggressive stand off, and moments later flippers were flying as a fight broke out.

Detail from the neck of an Emperor Penguin. These feathers overlay one another rather like roof tiles and reach densities of 12 feathers per sq cm (80 feathers per sq inch). They create a barrier that not even the harshest winter gales can penetrate.

When breeding Emperor Penguins come together after time apart they perform a mutual vocal display that helps to keeps the pair bond strong. It starts with both birds holding their heads high as they call, before drooping them in the fashion shown here. During an expedition to the Dawson-Lambton Glacier in Antarctica's Weddell Sea I spent many hours attempting to photograph this display. However I wanted an isolated pair where there were no other penguins as a distracting background. For a more intimate view I shot the image at penguin level, sitting on the ice just a short distance away.

Family life

During a spell camping at an Emperor Penguin rookery I switched my working schedule to sleep by day and photograph at night, when the sun was at its lowest. My diary entry for 12th November reads: 'I returned to camp at 2am to eat and warm up. The only food we seem to have left is bacon and maple syrup, an unlikely but very tasty combination. At 3am I ventured out again and made my way towards the head of the glacier. I soon came across an adult and chick standing quietly soaking up the warmth of the rising sun. It was around –3°C (27°F), but with no wind it felt quite balmy. I spent a wonderful hour with this penguin family, shooting into the light to use the backlit effects, which helped to create an enormously tranquil mood.'

An adult Gentoo Penguin responds to its chick's incessant begging and prepares to regurgitate partly digested food. This nest was located on a small plateau in a colony on Petermann Island on the Antarctic Peninsula, the southernmost Gentoo colony on Earth. The position allowed me to silhouette the scene against the setting sun, as the clock approached midnight.

An adult and chick Adélie Penguin enjoy the last rays of sun on a calm Antarctic evening. The French explorer Durmont d'Urville named this species after his wife Adéle.

This Adélie chick has just chased its parent around the colony to get a meal. Biologists think that these chases are designed to ensure a chick is fed away from the main flock. If a young bird took food in the presence of other chicks, then these others might jostle it in the scrum and the food could be needlessly spilt. If any of the meal falls on the ground it is automatically wasted, because penguins never take food that is soiled in this way.

It is not hard to see why young Emperors are used as characters in cartoons and animated movies.

A pair of Emperors bow to one another in courtship. It is late October and the chick is now growing fast, because the adults can return more frequently with food as the edge of the pack ice retreats towards the colony.

Born in July, the Emperor chick will be brooded on its parents' feet for 45–50 days. The adults also make overland feeding excursions to the open sea throughout August and September. These treks, which were made famous by the film *March of the Penguins*, may be 80km (50 miles) or more each way. During the five months that it takes for the chick to become independent, the parents may average around 16 feeding trips each to keep the young bird fed.

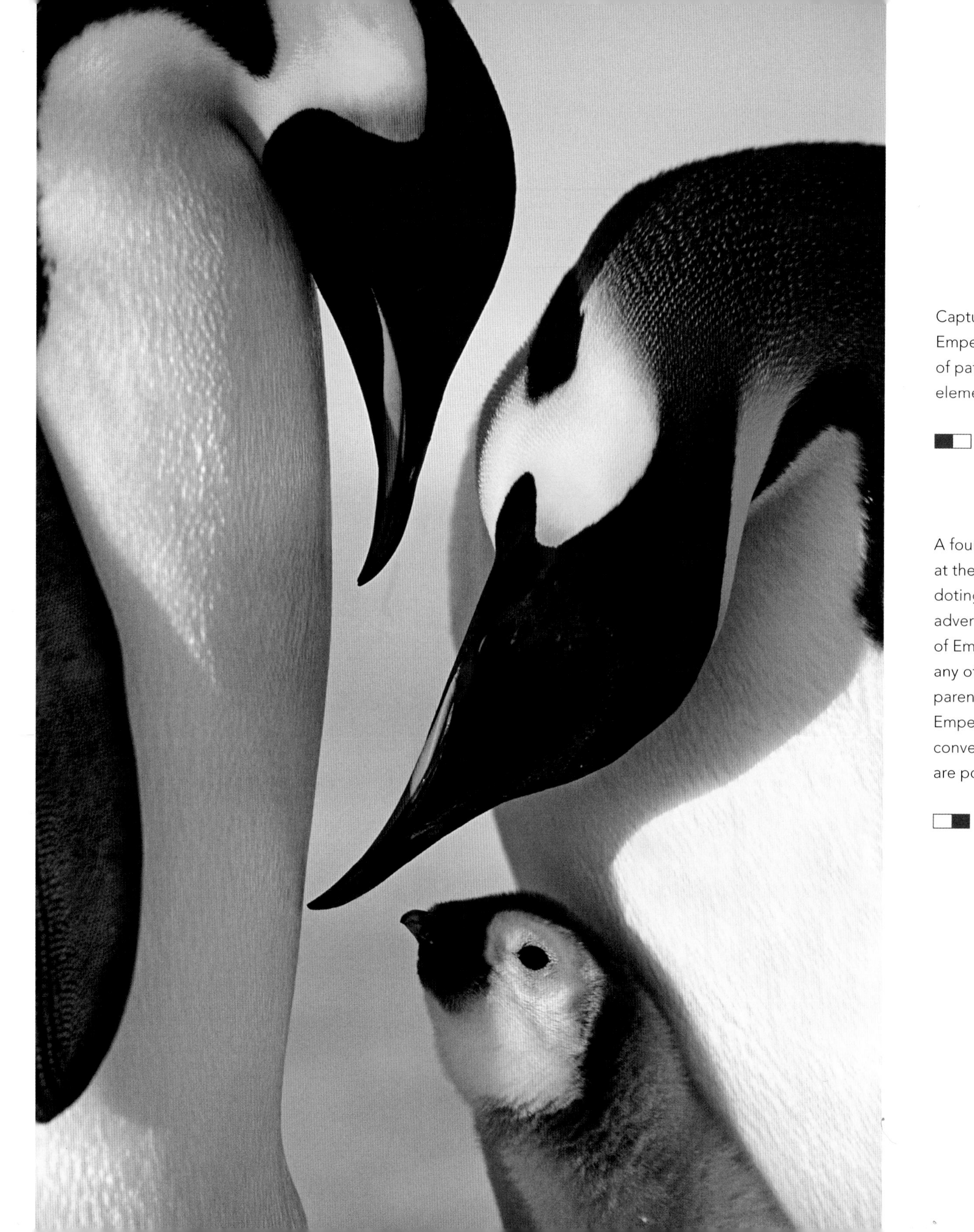

Capturing an image such as this of an Emperor family can take many hours of patient waiting until all the key elements of the image come together.

A four-month-old Emperor chick basks at the centre of attention from two doting parents. It is no surprise that advertising companies use images of Emperor Penguins more than any other bird pictures. Love, care, parental responsibility, risk – the Emperors' lifestyle can be used to convey so many of the qualities that are powerful drivers in our own lives.

Unlike most penguins, Emperors don't have a territory to defend. Instead, when incubating their egg or brooding their young, they shuffle around. Within a colony there may be many sub-colonies, groups standing together that change in size and structure. I lay on the ice next to this parent with its chick to capture an intimate portrait. There is surely no cuter baby bird on Earth than an Emperor that is just a few weeks old.

New life in one of Earth's harshest environments. The chick was so snuggled down and almost invisible in its parent's brood pouch that I had to wait for hours on end in conditions of –10°C (14°F) for its tiny head to appear. My long vigil was rewarded. The temperature in the adult's brood pouch is a constant 36° C (97°F). If tipped out accidentally on to the ice the chick would freeze to death in two minutes.

On the day this image was taken I was photographing in a T-shirt, the mercury having climbed to 18°C (65°F). The Emperor adults and chicks were panting and eating snow in order to cool down.

Standing like book ends in the Emperor colony at Snow Hill Island, these two chicks are still two months away from fledging and the moment when they finally head off to sea.

King Penguin chicks are looked after for around 50 weeks. Once the young birds reach adulthood they will not breed until they are at least 4–5 years old.

The brown fluffy chicks of King Penguins are often referred to as 'oakum boys'. This was the name given to them by the old sealers. Oakum boys once worked in shipyards and picked loose the old hemp ropes. These fibres were known as oakum and used to make the seams of ships watertight. The chicks reminded the sealers of handfuls of oakum.

In order to survive into their second year, King Penguin chicks must first make it through a South Georgian winter, having fasted and endured some of the worst weather on the planet. For such youngsters spring cannot come soon enough. Many will finally fledge in December, at the peak of the austral summer and a time of plenty in the Southern Ocean.

Walking back to my tent one morning, I came across this Emperor Penguin parent and chick on the edge of the colony. Successful wildlife photography is often about capturing a moment that conveys emotion to us. If that feeling readily translates as human sentiment, then the image can be all the more powerful.

It is impossible not to smile when in the company of penguins. The stern look of this adult Emperor at a playful chick could so easily be a scene from the film *Happy Feet*.

This image has a sinister twist. The chick is attempting to flee from adult Emperors that are intent on kidnapping it. Research suggests this is a strong biological urge in Emperors that is triggered by high levels of a hormone, prolactin. In normal circumstances the hormone ensures that parents are driven to return to the colony to attend their chick, after being away on their long foraging trips. However when they return to find that they have lost their own offspring, then other small unattended birds can become a target. A few minutes after this picture was taken the chick was dead, killed by the adults which, in their haste to collect and brood it on their feet, repeatedly fell on the unfortunate youngster.

Survival

My diary entry for the 17th November reads: 'The storm has raged for 24 hours now but last night we decided to try and make it to the colony. The three of us roped up and made our way over leads opening in the ice. At times there was no visibility and we had to stop and wait until we could see again. Eventually we reached the Emperors. Around 300 chicks were huddled together in a crèche. At times the colony, which I was right next to, would disappear from view in the blizzard. It was wild and very exciting to be in this storm with the birds.'

A lone Emperor chick is left out in the cold. When the temperature drops, the young huddle together in crèches to keep warm. This behaviour imitates that of their fathers who, when incubating the eggs in winter, cluster together and take turns to be on the outside. This formation is called a 'turtle'. Winds during winter can blow in excess of 160kph (100mph) and temperatures can drop below –20°C (–4°F) without the added effect of wind chill.

Penguins can have fierce fights and hit their opponents with their hard-edged flippers. The disputes are normally over mates or a nest site. These King Penguins squaring up to each other were on a beach on South Georgia.

A group of King Penguins bathes in the surf.

Emperor Penguin chicks huddle together to stay warm in the early hours of the morning.

Early in the breeding season penguins such as this Gentoo can be submerged in snow as they sit incubating their two eggs. They tend to nest on a plateau or on higher ground to avoid their nests being flooded during the thaw.

Antarctic Krill *Euphausia superba* swarm during summer in the Southern Ocean and are a staple food for penguins. In terms of biomass this small crustacean is probably the most abundant animal species on the planet. However, parts of Antarctica are some of the fastest-warming areas on Earth, and this has resulted in decreasing sea ice and a corresponding reduction in krill populations. The losses are having a knock-on effect on penguins, some species of which are in decline.

A Brown Skua snatches a Gentoo Penguin egg on Sea Lion Island on the Falklands. These opportunists need little encouragement to take eggs or chicks.

King Penguins on South Georgia put their backs to a blizzard. Winds exceeding 80kph (50mph) ripped through this colony as I crouched on its edge. Photographing in these conditions is difficult, with eyes constantly running and hands freezing even inside gloves.

King Penguins huddle together in a blizzard on South Georgia.

I found a viewpoint overlooking these King Penguins as they sheltered from strong katabatic winds. Often the whole group was obscured by blowing snow, but occasionally the wind would relent a little to allow a picture to be taken.

On a slope above Half Moon Bay on the Antarctic Peninsula, an Antarctic Skua circles a Gentoo Penguin as it sits guarding its eggs. Herbert Ponting the acclaimed photographer on Scott's British Antarctic expedition of 1910-12 called skuas 'the buccaneers of the South.' They can be very persistent and sometimes work in pairs to lure a penguin off its egg or away from its young. This skua eventually gave up and went off to try his luck elsewhere among the colony. The interaction made for an exciting photographic encounter.

Early spring and deep snow covers the route to and from a Gentoo Penguin colony. As they passed these two showed aggression towards each other, before continuing on their separate ways.

After hatching on their father's feet the young Emperor chick needs to be fed. If the female has not arrived by then, he feeds the chick with a milky substance from the lining of his oesophagus, a remarkable adaptation unique among penguins. Once the female returns she has to persuade him to pass the chick to her, while he goes to the sea to collect more food for the young bird. This changeover in the intense cold can be a risky moment in a young penguin's life and it has to be done extremely swiftly. Both adults face each other and the chick is spilled out on to the ice before being scooped up and brooded on the other partner's feet. If left on the ice for more than a couple of minutes the chick will die.

Some Leopard Seals take up station on the edge of penguin colonies and have an insatiable appetite for them. This seal has just killed an Adélie Penguin off Paulet Island. I spent an hour photographing it, in which time it caught six Adélie's very close in to the shore. Taking only small bites from each and tossing it about like a toy, the seal would soon turn his attention to catching another.

Penguins and us

Over the past four decades penguins have become major tourist attractions. Around 5,000 cruise-ship passengers visit South Georgia annually to enjoy the islands' wild bounty. Here a lone tourist looks over the King Penguin rookery at St Andrews Bay.

Penguins never fail to make us smile, whether it is because of their Chaplinesque walk or because of their inquisitive natures. In this case, the nosey parker is a juvenile King Penguin.

Studies have shown that the presence of researchers, scientists and tourists at penguin breeding colonies can have both good and bad consequences. Some species can become highly stressed, but others such as Magellanic Penguins might benefit from human company, because it deters natural predators from visiting the colony to take chicks or eggs.

Artists from a cruise ship paint at a King Penguin colony.

КАПИТАН ХЛЕБНИКОВ

QUARK EXPEDITIONS
FESCO

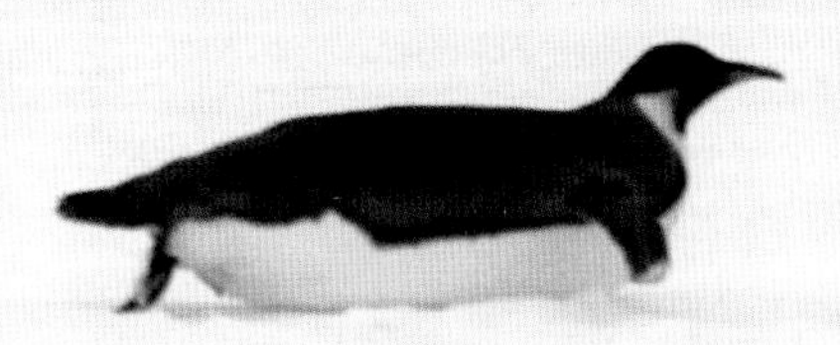

A small group of Emperor Penguins pass the *Kapitan Khlebnikov* as she sits in the pack ice of the Weddell Sea.

Penguins are popular exhibits in zoos and while few of us will have an opportunity to see the real thing in the wild, these captive birds at least give us a chance to get up close and personal with their fabulous antics. In this image, young children are enjoying African Penguins in a British zoo.

Penguins live in some of the most inaccessible places on the planet. Here in the beautiful Cooper Bay on South Georgia, photographers from a cruise ship sit quietly photographing a Macaroni Penguin colony.

A tourist photographs a colony of King Penguins.

As we watch penguins walking along with their upright gait and their flipper-like arms it is almost impossible not to compare them with ourselves. This group of King Penguins is strutting past a tourist at Salisbury Plain on South Georgia.

With no land-based predators in the Antarctic, other than skuas and giant petrels, Emperor Penguins show little fear of humans. Stand still for a few minutes on the edge of a colony and soon you will have curious chicks massing all around you.

While I was photographing at Snow Hill these Emperor chicks had sidled up to another person's camera. While he was away taking something from his bag the young birds lined up to pose.

Penguin fact file

Penguin taxonomy is hotly debated by different authorities, who recognize between 17 and 22 species. Some of these include the white-flippered form of Little Penguin *Eudyptula albosignata*. Rockhopper Penguins are also sometimes split into the Southern Rockhopper and Northern Rockhopper, while other ornithologists divide the Gentoo Penguin into two separate species. There are others that 'lump' the Royal Penguin with the Macaroni Penguin. In order to avoid confusion, I have listed only the 17 most widely recognized species.

Within the family there are six groups or genera. The brush-tailed penguins include the Gentoo, Chinstrap and Adélie Penguins. They have long tail feathers that look like a brush. The crested penguins are the six species with coloured head plumes and include those living around New Zealand – the Fiordland, Snares and Erect-crested Penguins – while the others are the Macaroni, Royal and Rockhopper Penguins. There are four species of banded or ringed penguins, which have ring-shaped bands of black and white on their chests and head. This quartet is also known as the warm-weather group and comprises the Galápagos, Humboldt, Magellanic and African Penguins. The King and Emperor Penguins, the largest of the family, are known as the great penguins. The remaining Little and Yellow-eyed are classified in their own separate groups.

The times I've spent photographing Emperor Penguins have been some of the happiest of my life. Although I liken it to living in a chest freezer, camping on the Antarctic ice is invigorating and provides opportunities for an extraordinary photographic harvest.

Emperor Penguin

Aptenodytes forsteri

Size: Largest of all the penguins. 112cm (44in), 27–41kg, (59.5–90lb).

Distribution: Circumpolar on Antarctic continent within limits of pack ice. All colonies except for two breed on sea-ice.

Prey: Fish, squid, and krill. Emperors are the deepest diving of all birds and have been recorded at depths in excess of 400m (1,300ft).

Breeding: Female lays egg then transfers to male to incubate for around 9 weeks through Antarctic winter on fast sea-ice. They huddle together to keep warm. Females time their return to coincide with the hatching of their chick, which is brooded on the adult's feet for the first few weeks of its life. Once the female has relieved the male he makes the journey across the ice to the open sea, which may be 80km (50 miles) or more away. He will have lost up to half his body weight during his long fast. The male returns after a few weeks. The parents then take it in turns to provide the chick with regurgitated meals.

Population: Latest population based on studying satellite images has resulted in an estimate of 238,000 pairs at 46 colonies. It is thought that Emperors are highly susceptible to climate change both warming and cooling.

King Penguin

Aptenodytes patagonicus

Size: 94cm (37in), 13.5–15kg (30–33lb).

Distribution: Breeds between 46°S and 55°S on subantarctic islands and peninsulas, usually forages widely in ice-free waters.

Prey: Primarily squid and fish.

Breeding: Breeds on bare ground and among vegetation close to the shore and on gently sloping hillsides, and along valleys behind beaches. Colonies range from a few dozen to hundreds of thousands. Elaborate courtship displays are followed by egg laying between November and April. Egg incubated on feet. After hatching the chick is brooded for around 6 weeks before both parents go off to forage leaving chick in crèche. Chicks fledge after 10–13 months. Owing to this long-cycle breeding regime, adults raise two chicks every three years.

Population: An estimated 1.5 to 2 million breeding pairs.

Adélie Penguin

Pygoscelis adeliae

Size: 46–61cm (18–24in), 3.5–4.5kg (7.7–10lb).

Distribution: Circumpolar on Antarctic continent within the limits of pack ice.

Prey: Primarily krill, also squid and fish caught in shallow dives, although deeper dives in excess of 150m (500ft) have been recorded.

Breeding: Breeding colonies are found on exposed rock around the Antarctic coast almost always near open water. Breed between October and April. Males arrive first and build nests of stones. Two eggs are laid incubated by both adults for 35 days. After hatching young tended for three weeks then crèche before fledging at two months. Young are susceptible to predation from skuas and giant petrels.

Population: Estimated 2 million breeding pairs. The population is increasing in East Antarctica and the Ross Sea, but decreasing on the Antarctic Peninsula, possibly because of the effects of global warming.
The loss of sea-ice cover is causing a decline in krill and small fish that are dependent upon ice formation.

Gentoo Penguin

Pygoscelis papua

Size: 70–80cm (27.5–31.5in), 5–7.5kg (11–16.5lb).

Distribution: Circumpolar in subantarctic and antarctic waters but avoids pack ice and continental coasts with the exception of the Antarctic Peninsula. Two subspecies recognised. *Pygoscelis papua papua* a larger form that breeds in the subantarctic on islands including the Falklands, South Georgia, Kerguelen, Heard, Macquarie and Staten. The smaller *Pygoscelis papua ellsworthi* breeds on Antarctic Peninsula, South Shetland, South Orkney and South Sandwich Islands.

Prey: Krill, squid and fish.

Breeding: Nests are made from stones and plant material. Two eggs are incubated for up to 39 days, young fledge after three months but can be fed by parents for a further period of up to 50 days. Adults are loyal to both nest sites and each other with long-lasting pair bonds common.

Population: Estimated at 314,000 pairs, with increases on Antarctic Peninsula and South Sandwich Island, but losses on some islands.

Chinstrap Penguin

Pygoscelis antarctica

Size: 50–72cm (19.5–28.5in), 4kg (8.8lb).

Distribution: Antarctic and subantarctic.

Prey: Primarily krill, but also small fish.

Breeding: Some very large colonies in excess of 100,000 pairs. Chinstraps choose ice-free areas and favour rocky slopes and headlands. Close bonds are formed with the same pair returning annually to the same nest site in October-November. Two eggs are laid in a shallow nest scrape surrounded by small stones. Both parents tend chicks equally and fledging occurs at 7–8 weeks old.

Population: About 8,000,000 individuals with range and population expanding.

Rockhopper Penguin

Eudyptes chrysocome

Size: 41–52cm (16–20.5in), 3kg (6.6lb).

Distribution: Widely on islands around the Southern Ocean.

Prey: Fish, octopus, squid and krill.

Breeding: Large colonies sometimes in excess of 100,000 pairs nesting on cliff tops and rocky gullies. Egg-laying commences between August and November dependent on location, with two eggs laid. Incubation for 33 days. After the chick hatches the male broods it for the first 25 days while the female makes feeding trips. Chicks then join crèches.

Population: Approx. 7,500,000 but some populations are in serious decline, with 90 per cent losses on the Falkland Islands over the past 60 years.

Royal Penguin

Eudyptes schlegeli

Size: 66–76cm (26–30in), 5.5kg (12lb).

Distribution: Restricted to Macquarie Island. Some authorities consider the Royal Penguin to be a subspecies of the Macaroni.

Prey: Primarily krill, squid and fish also taken.

Breeding: Birds come ashore in September. The nest is a scrape lined with grass and stones. Two eggs are laid in late October, but normally only the second egg is incubated fully and for around 35 days. Once hatched the chick is brooded by the male while the female visits with food for around 20 days. When old enough chicks form small crèches for another 50 days until fledging.

Population: Stable at about 850,000 breeding pairs.

Erect-crested Penguin

Eudyptes sclateri

Size: 62cm (24.5in), 2.5–4kg (5.5–8.8lb).

Distribution: Antipodes and Bounty Islands of New Zealand with smaller numbers on the Auckland and Campbell islands.

Prey: Thought to be krill, squid and small fish, but little is known of its prey choices.

Breeding: Return to breed in September in rocky areas with egg laying in October. Two eggs are laid in a nest of mud grass and stones. Just the second egg tends to be properly incubated for 35 days. On hatching the chick is tended for 35 days by the adult male, which fasts while the female brings food. Chicks fledge in February.

Population: Estimated 75,000 breeding pairs and in decline.

Macaroni Penguin

Eudyptes chrysolophus

Size: 71cm (28in), 5kg (11lb).

Distribution: Subantarctic islands in the Atlantic and Indian Oceans.

Prey: Krill and fish.

Breeding: Breeding colonies can be on rocky slopes or level ground and sometimes among tussock grass. Breeding commences in October/November. Two eggs are laid in a simple scrape, but the second is normally the only one that hatches. Incubation takes 37 days and once hatched the chick is brooded and guarded by the male while the female brings food. Older chicks then form crèches and fledge about 70 days after hatching.

Population: 9 million breeding pairs. Discussion exists over apparent declines in population. These may have resulted from changes in survey methods, although rapid declines have now been reported at a number of localities including South Georgia.

Fiordland Penguin

Eudyptes pachyrhynchus

Size: 55cm (21.5in), 3–4kg (6.6–8.8lb).

Distribution: Breeds on the south-western coast of South Island in New Zealand and on nearby Stewart and Solander Islands.

Prey: Small fish, crustaceans and squid.

Breeding: Pairs breed among rocks or burrows and cavities among tree roots in coastal forest. Two eggs are laid in July or August, but usually just one chick survives. Normally the male then looks after the chick until it is old enough to be left. Chick fledges after 10–11 weeks.

Population: Estimated 2,500 pairs, declining under pressure from predation and human interference.

Snares Penguin

Eudyptes robustus

Size: 60–70cm (23.5–27.5in), 4–7kg (8.8–15.5lb).

Distribution: South of New Zealand in the Snares Islands.

Prey: Krill, fish and squid.

Breeding: Colonies found in wooded areas where nest is a shallow dug hole lined with twigs under a shrub or tree to afford shade. Two eggs laid September-October, but normally only one chick survives. After hatching the chick is guarded by the male while the female returns to feed chick. Young fledge after 11 weeks.

Population: About 46,500 adult penguins. All occur in fewer than five locations, but current population probably stable.

Yellow-eyed Penguin

Megodyptes antipodes

Size: 65–68cm (25.5–26.8in), 5–8kg (11–17.6lb).

Distribution: New Zealand including south-east coast of South Island, Foveaux Strait, Auckland, Campbell and Stewart Islands.

Prey: Fish.

Breeding: Breeds in forest and scrub nesting against trees or rocks. Eggs are laid in September or October in a nest of sticks, incubation can take two months and chicks fledge after 60–70 days.

Population: 2,000 breeding pairs making this species the second rarest penguin in the world.

Magellanic Penguin

Spheniscus magellanicus

Size: 61–71cm (24–28in), 4–5kg (8.8–11lb).

Distribution: South America, including the coasts of Chile, Argentina and the Falkland Islands.

Prey: Small fish particularly anchovies, squid and crustaceans.

Breeding: Nests in grassland and on coasts and cliff faces. Sometimes the birds make a simple scrape under a bush but, where ground conditions permit, they often use a burrow. They start nesting in September or October, laying two eggs incubated for around 40 days. The chicks are brooded for around 25–30 days, and after that they are left unattended and visited every one to three days. Young fledge between January and March.

Population: Around 1.3 million pairs.

African Penguin

Spheniscus demersus

Alternative names: Jackass Penguin, Black-footed Penguin.

Size: 61–71cm (24–28in), 3kg (6.6lb).

Distribution: Southern Africa.

Prey: Fish, mostly anchovies and sardines. Also squid and crustaceans.

Breeding: No fixed breeding season, but peaks in Namibia between November and December and in South Africa between March and May. Nests in scrapes under bushes and boulders, or in burrows. Incubation of two eggs takes 40 days and chicks will be guarded for 30 days before forming a crèche. Young fledge at between 60 and 130 days old.

Population: Fewer than 25,000 breeding pairs. This species has been in rapid decline partly because of overfishing of anchovies and sardines and redistribution of food in the ocean.

Little Penguin

Eudyptula minor

Alternative names: Little Blue Penguin, Blue Penguin, Fairy Penguin.

Size: 41cm (16in), about 1kg (2.2lb).

Distribution: Southern Australia and New Zealand.

Prey: Small fish and squid.

Breeding: Nests in burrows, caves, crevices and under bushes in colonies or singly in pairs. Breeding season is variable depending on location but commonly starts in August or September with two eggs laid. Chicks fledge in November or December. Second and third clutches can be laid, extending the breeding season until May. Parents leave chicks at dawn to fish all day and return soon after sunset.

Population: 250,000 breeding pairs.

Humboldt Penguin

Spheniscus humboldti

Size: 56–66cm (22–26in), 4kg (8.8lb).

Distribution: Found on islands off western South America, and along the coasts of Peru and Chile within the Humboldt Current.

Prey: Anchovies and other fish species, also squid and krill.

Breeding: Humboldt Penguins hang around breeding colonies on rocky coasts and islands throughout the year with two main breeding seasons – March to April or September to October – depending on location. Nest in burrows or crevices laying two eggs with incubation taking around 40 days, fledging occurs after 12 weeks.

Population: Estimated to be 3,300 to 12,000 breeding adults. This species faces severe threats including over-fishing of prey species, drowning in fishing nets, guano harvesting, human interference, and El Niño Southern Oscillation events, which have had a major impact on numbers in the last two decades.

Galápagos Penguin

Spheniscus mendiculus

Size: 53cm (21in), 2.5kg (5.5lb).

Distribution: Galápagos Islands primarily on Fernandina and Isabela, which are fed by the cold Cromwell Current.

Prey: Small fish.

Breeding: Pairs bond for life, two eggs are laid and incubated for 40 days in a simple nest in shade or in a rock crevice. Chicks fledge after 65 days. Breeding occurs when feeding conditions are good.

Population: Estimated 500 breeding pairs. This is the world's most endangered penguin. The major threat is El Niño Southern Oscillation events, which can have a heavy impact on population.

Acknowledgments

Flying into Antarctica and camping next to an Emperor Penguin colony requires a great deal of logistical support. For this I thank John Noble and the Adventure Network team at Patriot Hills in 1998, including mountaineer Conrad Anker who kept us safe on the ice.

Cheesemans' Ecology Safaris made many of the pictures within this book possible by ensuring safe shore landings in often less than favorable conditions. I thank fellow photographer Roger Tidman for his camaraderie on a number of trips to the far south. My thanks to Mark Cocker for commenting on the text.

My family, Jayne, James and Charlotte have given their unwavering support despite my frequent absences in pursuit of penguins. I owe them so much.

References

Deguine, J-C. 1974. ***Emperor Penguin.*** Stephen Greene Press, Vermont.

Jacquet, L. and Roberts, J. 2006. ***The March of the Penguins.*** National Geographic Society, Washington.

Lynch, W. 2006. ***Penguins of the World.*** A&C Black, London.

Muller-Schwarze, D. 1984. ***The Behaviour of Penguins.*** State University of New York Press, New York.

Polking, F. 2006. ***Penguins.*** Evans Mitchell, London.

Shirihai, H. and Jarrett, B. 2002. ***A Complete Guide to Antarctic Wildlife.*** Alula Press, Finland.

Williams, T., Davies, J.N. and Busby, J. 1995. ***The Penguins.*** Oxford University Press, Oxford.

Wilson, E. 1966. ***Diary of the Discovery Expedition.*** Blandford Press, London.

Index